CLIMATE CHANGE

# CLIMATE CHANGE

ODYSSEYS

JESSICA GUNDERSON

CREATIVE EDUCATION · CREATIVE PAPERBACKS

Published by Creative Education and Creative Paperbacks
P.O. Box 227, Mankato, Minnesota 56002
Creative Education and Creative Paperbacks are imprints of
The Creative Company
www.thecreativecompany.us

Design by Blue Design; production by Colin O'Dea
Art direction by Rita Marshall

Photographs by 123RF (Mark Agnor), Alamy (AF archive, Dinodia Photos), Creative Commons Wikimedia (Jessica Hochreiter/Arizona State University, Charles H. Smith/U.S. Fish and Wildlife Service), Getty Images (Anton Petrus/Moment, Liyao Xie/Moment), iStockphoto (AlessandroPhoto, amriphoto, brainmaster, Kevin Brine, carlosrojas20, kev303, Kwanchai_Khammuean, MediaProduction, Nazarli Neshcherenskyi, RelaxFoto.de, Florent Rols, saschasid, shaunl, SimonSkafar, spxChrome, Surachet1, travellinglight, urfinguss), National Geographic Creative (MARK THIESSEN), Shutterstock (Otmar Smit)



Library of Congress Cataloging-in-Publication Data
Names: Gunderson, Jessica, author.
Title: Climate change / Jessica Gunderson.
Series: Odysseys in the environment.
Includes index.
Summary: An in-depth study of climate change, examining historical and present-day causes and effects of rising global temperatures, as well as what people can do to combat climate change.
Identifiers: ISBN 978-1-64026-426-7 (hardcover) / ISBN 978-1-62832-959-9 (pbk) / ISBN 978-1-64000-600-3 (eBook)
This title has been submitted for CIP processing under LCCN 2021937897.

# CONTENTS

**Introduction** . . . . . . . . . . . . . . . . . . . . . . . . **9**

**An Atmosphere of Life** . . . . . . . . . . . . . . . . . . **13**

Advocating for Action . . . . . . . . . . . . . . . . . . . . 19

Water in the Atmosphere . . . . . . . . . . . . . . . . . . 22

**Emitting the Problem** . . . . . . . . . . . . . . . . . . . **29**

Capturing Carbon . . . . . . . . . . . . . . . . . . . . . . 33

The Ozone Hole . . . . . . . . . . . . . . . . . . . . . . . 36

The Death of Coral Reefs . . . . . . . . . . . . . . . . . . 40

**Warming Effects** . . . . . . . . . . . . . . . . . . . . . **45**

Golden Toad Extinction . . . . . . . . . . . . . . . . . . 48

Inconvenient Truths . . . . . . . . . . . . . . . . . . . . 50

Cleaning Fossil Fuels . . . . . . . . . . . . . . . . . . . 54

**A Worldwide Effort** . . . . . . . . . . . . . . . . . . . **59**

Alternate Energy . . . . . . . . . . . . . . . . . . . . . . 62

Shrinking Footprints . . . . . . . . . . . . . . . . . . . . 66

**Selected Bibliography** . . . . . . . . . . . . . . . . . . **74**

**Endnotes** . . . . . . . . . . . . . . . . . . . . . . . . . **75**

**Websites** . . . . . . . . . . . . . . . . . . . . . . . . . **78**

**Index** . . . . . . . . . . . . . . . . . . . . . . . . . . . **80**

# Introduction

Everything human beings need to survive—air to breathe, food to eat, water to drink—is found on Earth, and on Earth alone. Yet the very planet that sustains human life has come under threat because of human activities. Rivers and lakes are drying up as people divert water for their own use. Temperatures are rising as greenhouse gases such as carbon dioxide trap heat in the **atmosphere**. Species of plants and animals are

**OPPOSITE:** **The world's lakes are gradually warming, causing an increase in algae and bacteria harmful to fish and people.**

disappearing as people destroy essential habitats. And the rate of many such changes appears to be accelerating. "If I had to use one word to describe the environmental state of the planet right now, I think I would say precarious," said population expert Robert Engelman. "It isn't doomed. It isn't certainly headed toward disaster. But it's in a very precarious situation right now."

One of the biggest threats facing Earth is global warming. The average global temperature is steadily rising, causing polar ice to melt, sea levels to rise, and strong storms to brew. Most scientists believe that the rising temperatures are partially caused by human actions and could bring on mass extinctions and other disasters. As Al Gore, former vice president of the United States and 2007 Nobel Peace Prize winner, warned, "We will face a string of terrible catastrophes unless we act to prepare ourselves and deal with the underlying causes of global warming." But what exactly are the causes? Are humans really at fault? And what can we do to stop global warming before life on Earth is destroyed?

# An Atmosphere of Life

Earth's climate—the long-term weather conditions throughout the world over a period of time—is constantly changing. Over the planet's 4.6 billion years of existence, many natural changes in climate have occurred. For instance, ice ages, or periods in which large portions of Earth have been covered with sheets of ice, have occurred at regular intervals for the past two million years.

---

**OPPOSITE:** Earth's five warmest years on record have all occurred since 2015, according to the National Oceanic and Atmospheric Administration (NOAA).

TAKEAWAY

Although many scientists believe that global warming is natural to an extent, most agree that human activity has quickened its pace.

Between ice ages, warming spells called interglacial periods caused the widespread snow and ice to melt and created the conditions that sustain life on the planet today. Climate scientists, called climatologists, study such changes and have been keeping detailed records of Earth's climate for the past 150 years. What they have seen in the past three decades has been cause for alarm: Earth's temperature is rapidly increasing and shows no signs of stopping. Although many scientists believe that global warming is natural to an extent, most agree that human activity has quickened its pace.

Life on Earth is sustained in large part thanks to the

protective nature of the atmosphere. Stretching from the surface of the earth to about 600 miles (966 km) into space, the atmosphere is held close to the earth because of gravity. It is divided into four successive layers, which, from bottom to top, are the troposphere, stratosphere, mesosphere, and ionosphere, or thermosphere. The atmosphere is made up primarily of a colorless, odorless gas called nitrogen. The remainder (about 22 percent) is mostly oxygen, with small amounts of water vapor, carbon dioxide, and several other gases.

The combination of gases in the troposphere make up what is known as air. Air is essential to life. Humans, animals, and plants use elements found in the air to change food into energy. Two critical elements in this conversion process are oxygen and carbon, and the movement of these elements within and between the

**OPPOSITE** The element carbon is a part of the atmosphere, seawater, soil, rocks, and all living things.

air, the earth, and all living things is described by the oxygen and carbon cycles. A process called photosynthesis is critical to both cycles. In photosynthesis, plants absorb sunlight and use that energy to combine carbon gas, or carbon dioxide, with water to produce sugars. In the process, water molecules are split, which releases oxygen into the atmosphere. Animals, on the other hand, breathe in oxygen and exhale carbon dioxide during **respiration**. Thus, the carbon and oxygen cycles work together to maintain a balance of oxygen and carbon in the atmosphere.

The carbon cycle circulates carbon by photosynthesis and respiration, as well as by other means. Plants contain large amounts of carbon, so when they burn or decay, they release carbon dioxide into the air. Over millions of years, dead plants beneath the earth's surface compact

**OPPOSITE** Researchers estimate that roughly 20 percent of the planet's original forests remain intact.

and become fossil fuels such as oil, coal, and natural gas. Burning fossil fuels releases carbon dioxide into the atmosphere. Oceans, which absorb vast amounts of carbon dioxide, release some of it back into the air and dissolve the rest into carbonic acid. This cyclical exchange of carbon between land, oceans, and the atmosphere is vital to all plant and animal life.

The atmosphere not only contains gases that sustain life, but it also helps keep Earth at a moderate temperature that allows plants and animals to thrive. This is called the greenhouse effect. The greenhouse effect was first proposed in 1827 by French scientist Jean-Baptiste Fourier, who compared Earth's atmosphere to the glass of a greenhouse. The sun's warmth penetrates the glass structure of the greenhouse, warming the air inside. When the sun goes down, the glass traps the heat inside, keeping

# Advocating for Action

The Union of Concerned Scientists (UCS) is an American nonprofit organization aimed at protecting the environment and making the world a safer place in which to live. The UCS was founded in 1969 and now has more than 500,000 supporters who are united in their concern for the planet's health. The organization uses both independent research and citizen action to persuade governments to adopt changes and solutions to environmental issues. Climate change is one of the organization's top priorities, and one of its most recent successes was bringing deforestation to the forefront of climate-based discussions.

the air warm even without the sun's rays. Greenhouse gases in the atmosphere operate in a similar way. The gases form a blanket around Earth, keeping the planet's temperature at an average of 57 °F (14 °C). Without the greenhouse effect, Earth's average temperature would be about 0 °F (-17.8 °C), much too cold to sustain life.

The most abundant greenhouse gases are water vapor, carbon dioxide, methane, nitrous oxide, and **ozone**. For thousands of years, the balance between these gases remained the same. However, over the past 150 years, humans have released more and more of these gases into the air. Since the **Industrial Revolution**, humans have increasingly relied upon machines for transportation, heating and cooling, and manufacturing goods. These machines are often powered by the burning of fossil fuels, which releases carbon dioxide into the atmosphere.

TAKEAWAY

Since the beginning of the Industrial Revolution, the amount of carbon dioxide in the atmosphere has increased by an estimated 31 percent.

In 1905, Swedish scientist Svante Arrhenius suggested that the carbon emissions from burning fuels could cause problems for the planet. As humans burned more fossil fuels, Arrhenius warned, the carbon cycle's natural balance would be upset. He predicted that the presence of a higher level of greenhouse gases in the atmosphere could cause the planet to warm.

Carbon dioxide is the most prevalent greenhouse gas emitted by human activities. Since the beginning of the Industrial Revolution, the amount of carbon dioxide in the atmosphere has increased by an estimated 31

## Water in the Atmosphere

Water vapor is the most abundant greenhouse gas and the biggest contributor to the natural greenhouse effect. However, because the amount of water in the atmosphere—whether a vapor (gas), liquid, or solid—never changes, it does not contribute to the enhanced greenhouse effect. The concentration of water vapor in the air, known as humidity, can change, however. High levels of humidity increase warming because water vapor, as a greenhouse gas, keeps heat from escaping. But high levels of water vapor can actually help cool the earth, too, by causing clouds to form. Clouds reflect the sun's energy away from the earth, producing a cooling effect.

TAKEAWAY

Scientists estimate that carbon dioxide accounts for nearly a quarter of all global warming.

percent. For many years, scientists believed that natural environments known as carbon sinks, such as forests and oceans, would always be able to soak up the excess carbon dioxide. However, in the past few decades, humans have been generating more carbon dioxide than these sinks can absorb. Excess carbon dioxide is then released into the atmosphere, making the greenhouse gases' blanket around Earth thicker and warming the globe in what is known as the enhanced greenhouse effect. Scientists estimate that carbon dioxide accounts for nearly a quarter of all global warming.

To fully understand the effects humans have had on the earth, scientists must study the planet's climate history. They can closely estimate Earth's prior climate patterns by using various methods of dating certain natural objects such as trees and ice. Scientists study tree rings by slicing open a section of an old tree. Each year, trees grow a new layer of bark, creating a ring around the older layers. During wet years, the ring is thicker, so examining the thickness of the ring helps scientists estimate the precipitation and other climatic factors of that year. Climatologists also study ice cores. Some ice in polar **latitudes** is hundreds to thousands of years old and contains pockets of gas that are trapped within it. Drilling into the ice and studying the gas trapped in these pockets helps scientists determine the makeup of the atmosphere during that time period. Further, the ice core studies have shown a direct **correlation** between higher levels of carbon dioxide in the air and an increased

**OPPOSITE** In some areas, climate change has caused soil to become drier, which leads to smaller crop yields and the need for more irrigation.

average temperature of the planet.

Using the data obtained by studying Earth's climate history, scientists have been able to determine that in the last 100 years, Earth's average temperature has risen 1 °F (0.6 °C). Although that number may seem insignificant, even a small increase in temperature has a large impact on the planet, causing polar ice caps to melt, sea levels to rise, and droughts, storms, and floods to occur with greater frequency. Scientists warn that average global temperatures may increase by up to 7.2 °F (4.3 °C) by the year 2100, in large part because of the continued release of greenhouse gases into the atmosphere.

# Emitting the Problem

Most scientists believe that humans have caused Earth's rapid climate change. Human activities, especially burning fuel for energy, have emitted excessive amounts of greenhouse gases into the air; in the 1990s alone, nearly 8 billion tons (7.3 billion t) of carbon dioxide were released each year. Despite the pandemic-induced plunge in global emissions in 2020, the International Energy Agency

**OPPOSITE:** **Exhaust goes into the air and creates air pollution called smog, which looks like a smoky fog.**

TAKEAWAY

Coal and natural gas are used for electricity and heating, while oil is used for transportation.

warned that emissions could rebound dangerously.

Because people in most regions of the world rely on fossil fuels to supply their energy needs, the carbon-rich fuels account for the majority of the harmful emissions. Coal and natural gas are used for electricity and heating, while oil is used for transportation. Cars and trucks run on gasoline, which is a flammable liquid made from **crude oil**. Every day, Americans burn 389.89 million gallons, according to the U.S. Energy Information Administration. The U.S. uses about 20 percent of the world's oil supply, and most of it goes toward gasoline. Some **developing countries**, such as China, also use large

**PREVIOUS SPREAD** Motor vehicles are among the largest contributors of climate change pollution in the world.

amounts of fossil fuels. China surpassed the U.S. in oil imports in 2017.

Carbon sinks no longer have the capacity to absorb all of the excess carbon dioxide being released by burning such fuels, and to add to the problem, many such sinks are disappearing. In the oceans, **phytoplankton** absorb large amounts of carbon dioxide for use in photosynthesis. But in recent years, ocean pollution has been killing phytoplankton, and more carbon dioxide therefore goes into the atmosphere.

## Capturing Carbon

In 1999, an American eighth grader named Claire Lackner conducted a science experiment that used a fish tank and pump to pull carbon dioxide from the air. Her project gave her father, Klaus (pictured), founder and director of the Center for Negative Carbon Emissions at Arizona State University, an idea. He developed a **synthetic** tree that captures carbon dioxide from the air. More of a process than machines at present, the "trees" are about 1,000 times more effective than living trees. In a year, one synthetic tree can absorb and store the carbon dioxide emissions equivalent to 7,300 cars. As of 2022, this new technology was growing closer to becoming a reality.

TAKEAWAY

In some parts of the world, forests are cleared, and then the land is used for slash-and-burn agriculture.

Deforestation—removing vegetation to make room for buildings, roads, and farmland—reduces the supply of available sinks on land. Worldwide, an area the size of one or two football fields is lost each second from deforestation. Industrial logging removes scores of trees from the world's forests as trees are chopped down to make building materials and paper or for use as fuel. Many logging companies plant seedlings in place of the cleared trees; however, the small trees may take many years to grow large enough to absorb a significant amount of carbon dioxide.

In addition to removing valuable carbon sinks, deforestation also increases atmospheric carbon dioxide directly. In some parts of the world, forests are cleared, and then the land is used for slash-and-burn agriculture. First, trees are chopped down and removed, and the tree stumps that remain are burned. The ash from the burned stumps adds nutrients to the soil, making it good for short-term farming, but the act of burning turns the carbon stored inside the stumps into carbon dioxide, releasing it directly into the atmosphere.

Carbon dioxide is not the only greenhouse gas responsible for global warming. Methane, which can also occur naturally, is the second-most abundant greenhouse gas that is increasing because of human activity. Methane traps 20 times more heat than carbon dioxide and is a major concern to climatologists. Landfills are the main culprits of methane emissions because, as waste decom-

# The Ozone Hole

In the 1970s, scientists discovered that the ozone layer in the upper atmosphere was thinning, and in 1985, an "ozone hole" was found in the stratosphere over Antarctica. People often believe that the ozone hole and global warming are related; however, the ozone hole was not caused by global warming. Instead, the ozone was depleted in large part by human-made compounds commonly used in spray cans. In the upper atmosphere, ozone intercepts the sun's harmful ultraviolet (UV) rays, protecting humans from skin cancer, eye damage, and other problems. These compounds have been banned worldwide. The ozone hole is on the mend.

TAKEAWAY

Nitrous oxide levels have increased 18 percent since the beginning of the Industrial Revolution.

poses, or breaks down, methane is released. Another source of methane comes from grazing livestock such as cows. When cows digest food, they release methane as a byproduct. In one day, a cow can release anywhere from 50 to 150 gallons (189–568 l) of methane. Rice fields also produce methane. Rice is grown on flooded fields. Because the soil is so wet, it cannot receive oxygen. So, the **organic** matter present in the soil breaks down and releases methane. Additionally, when people drill for oil, the drills cut into methane pockets found in the earth, releasing methane into the air.

Nitrous oxide is another greenhouse gas that humans generate. Car exhaust, factory smokestacks, and fertilizers used on farms all produce nitrous oxide. Nitrous oxide levels have increased 18 percent since the beginning of the Industrial Revolution. Human-made chemicals also contribute to the enhanced greenhouse effect. Chemical compounds known as chlorofluorocarbons (CFCs) were developed in the 1930s. They were commonly used as coolants in refrigerators, air conditioners, and freezers until the 1970s, when scientists realized the dangers of the gas. CFCs remain in the atmosphere for up to 100 years and trap more heat than any other greenhouse gas.

The world's increasing population is putting a strain on available resources and adding more greenhouse gases to the atmosphere. As of 2021, the world population was

# The Death of Coral Reefs

Coral reefs are complex underwater structures built from the hard outer skeletons of individual animals called coral polyps. Found in tropical waters, coral reefs are important to marine life; one in four sea creatures spends at least part of its life among coral reefs. Coral reefs also provide natural barriers that protect many tropical coastlines from the full force of the ocean's waves. However, the reefs are suffering damage from climate change, mostly from coral bleaching. Corals get their color and food from algae that grow near them, but warmer ocean temperatures have caused algae to disappear. The corals then "bleach," or turn white and die.

7.9 billion. As the population increases, more people will need fossil fuels for heating and transportation. More crops for food will need to be grown, and the machines that harvest those crops will emit more greenhouse gases. In addition, more fertilizers will be used to enrich the soil, releasing nitrous oxide into the air and into nearby water supplies as well.

Along with a rising population, the world is witnessing a growing middle class. More people now own their own cars. They also often live in single-family houses and own products such as smart phones, computers, and televisions. Natural resources are needed to manufacture these electronic products, and energy is needed to use them. Energy use emits greenhouse gases, drains the earth of its natural fossil fuels, and creates waste.

The invention of cars, airplanes, and trains has made

**OPPOSITE** In 2017, airplanes around the world took to the air for an average of 102,465 flights each day.

transportation easier for almost all people. People are able to travel from place to place quickly, and many people commute several miles to get to their workplaces. People also travel to visit friends and family or to go on vacation. The commuter lifestyle has contributed to the emission of greenhouse gases, as car exhaust releases nitrous oxide and carbon gases into the air, and travel by train and airplane uses fossil fuels as well. Some experts say air travel is particularly damaging to the environment because the airplanes emit gases and other pollutants directly into the upper troposphere and lower stratosphere.

# Warming Effects

Some debate lingers over whether high levels of greenhouse gases directly cause global warming, but the evidence is clear that warming is occurring. Areas near the poles are warming faster than the rest of the globe. Over the past 50 years, temperatures have risen about 7 °F (4 °C) in the Arctic and 4.5 °F (2.5 °C) in the Antarctic. These warmer temperatures have caused age-old snow and ice to melt. For instance, the

**OPPOSITE:** Researchers suggest that the Arctic is warming twice as fast as anywhere else on Earth, and every 10 years, Arctic sea ice shrinks by 10 percent.

Larsen B Ice Shelf, a 12,000-year-old area of Antarctic ice, broke up in a matter of months in 2002. But large-scale melting is not restricted to polar regions; Montana's Glacier National Park contained 150 glaciers in 1910. In 2017, 26 remained. The U.S. government estimates that there will be no glaciers remaining in the national park by the year 2030.

The melting ice caps, snow, and glaciers heighten the effects of global warming. Because ice and snow are white, they reflect about 90 percent of the light from the sun. The percentage of reflected sunlight is called albedo. When an area warms and the snow or ice melts, it exposes darker land or sea. Darker colors are less reflective and absorb more heat, causing Earth's temperature to rise. The higher temperatures cause more ice and snow to melt, and the cycle continues in what is called

TAKEAWAY

Higher sea levels can endanger coastal communities around the world.

the albedo effect.

Melting snow and ice also affect sea levels. As the snow and ice on land melt, water runs into lakes and rivers, which flow into the oceans, and the ice already present in oceans melts directly into the water. Even without the melting sea ice, though, sea levels are rising because of higher global temperatures that have caused entire bodies of water to warm and expand. Higher sea levels can endanger coastal communities around the world. Seaside cities such as New Orleans, Louisiana; Rio de Janeiro, Brazil; and Shanghai, China, would face certain flooding, and low-lying countries such as the

# Golden Toad Extinction

The golden toad was the first known species to disappear because of global warming. The brilliantly colored toad once lived exclusively in Costa Rica's Monteverde Cloud Forest Reserve. Cloud forests are characterized by an almost constant cloud cover or fog that hovers above the trees, and the golden toad needed these conditions to breed. However, warming of the surrounding ocean caused the fog to form farther up the mountain, robbing the toad of its habitat and making it more susceptible to disease and direct heat from the sun. First identified in 1966, no one has seen a golden toad since 1989, and the species is considered extinct.

Netherlands, Bangladesh, and the island nation of the Maldives are concerned about losing their inhabitants, **economic** independence, and entire way of life. The Netherlands has built **dikes** and drainage systems to combat the rising waters, but other countries in similar straits do not have the physical or financial resources to undertake such projects.

Meanwhile, freshwater supplies are being hurt by rising sea levels. When salt water from the oceans spills over into freshwater lakes and rivers, the water used for drinking and **irrigation** becomes contaminated. The proportion of fresh water available to the amount withdrawn is called water stress. High water stress indicates that there is a small ratio of supply to the amount needed. In 2018, the United Nations reported that more than 2 billion people lived in countries experiencing high water stress.

## Inconvenient Truths

Former U.S. vice president Al Gore is an influential **advocate** of the movement to reduce global warming. He first became interested in environmental studies as a college student at Harvard. He remained involved in environmental issues throughout his life and career in politics, publishing his first book, *Earth in the Balance*, in 1992. His documentary film and companion book, *An Inconvenient Truth*, directly address the consequences of and solutions to global warming. The documentary won an Academy Award in 2007 and is considered one of the most influential documentaries of the 21st century. For his work on environmental issues, Gore received the 2007 Nobel Peace Prize.

The warming oceans contribute to changing weather patterns. As an ocean's surface warms, water evaporates, which increases the energy in the atmosphere. The heightened energy in the atmosphere can produce more instances of extreme weather, such as hurricanes and tornadoes. In 2005, the earth experienced the most catastrophic storms on record (until 2020, which surpassed 2005's total). In August of that year, Hurricane Katrina ravaged the coast along the Gulf of Mexico, killing more than 1,300 people and displacing millions. In 2017, hurricanes Harvey, Irma, and Maria washed across the Caribbean and parts of the U.S. within about a six-week time period. Although global warming is not the only cause of such extreme weather, many climatologists agree that the overall warming temperatures contribute to such storms.

Some areas of the globe have recently experienced heat waves that lasted for days or even months. Many areas are not used to such high temperatures, and people often do not have the resources, such as air conditioning, necessary to combat the heat. In 2003, a heat wave in Europe caused at least 30,000 deaths.

Rising global temperatures also contribute to desertification, or the process by which land is turned into desert. High heat combined with low levels of precipitation can cause desertification, as can droughts, which are also becoming increasingly more common. A drought is a period with little to no precipitation. Droughts can destroy food crops and water supplies. In less-wealthy countries that depend heavily upon agriculture, failed crops can mean widespread starvation. These countries may not be able to feed their people if the food supply is low.

## Cleaning Fossil Fuels

One of the main reasons people use fossil fuels is because they are cheap forms of energy. Environmentalists worry that most nations and factories will choose the cheaper energy sources rather than investing in more Earth-friendly alternatives. They argue that, at the very least, we need to make fossil fuels cleaner, and one way to do this is by using clean coal technology at coal-fired power plants. Clean coal technology captures carbon dioxide before it is released into the air, then pumps it underground for storage. Implementing clean coal technology is expensive, however, and experts say it is decades away from being perfected for widespread use.

Climate change affects plant and animal species around the world, most notably the polar bear. Polar bears travel across permanent sea ice and use floating ice to hunt for seals, but because of rapid melting, their hunting options are shrinking. In addition, the ice that forms every winter is melting sooner than usual, sometimes stranding the bears, which are then unable to **migrate** to summer hunting grounds. Extended periods without food have caused their populations to dwindle, and in 2007, the U.S. Geological Survey warned that if polar ice continues to melt, two-thirds of all polar bears could disappear by 2050.

In other areas, as the temperatures increase, cold-weather species are moving away to find cooler habitats, living closer to the poles or climbing to higher **altitudes**. Birds' migratory patterns are changing, with

**OPPOSITE** As temperatures rise, so do the chances that more people will suffer from mosquito-borne diseases, including malaria, Zika, and dengue.

some birds ceasing to migrate at all because of increasingly milder winters. Species that are unable to adapt to the changing climate are at risk of becoming extinct, and extinct species can upset the natural balance of a habitat's **food chain**.

Human health, too, is affected by climate changes when diseases such as **malaria** spread to areas in which they previously did not exist. Malaria is spread by infected mosquitoes and is found mostly in tropical regions surrounding the equator. However, as temperatures warm throughout the globe, infected mosquitoes are spreading to nontropical areas. Higher temperatures also affect the quality of the air people breathe, especially in crowded cities, as warmer air causes smog (a mixture of fog, smoke, and other air pollutants) to linger. High levels of smog can bring on chest pains and asthma attacks.

NO
PLANET

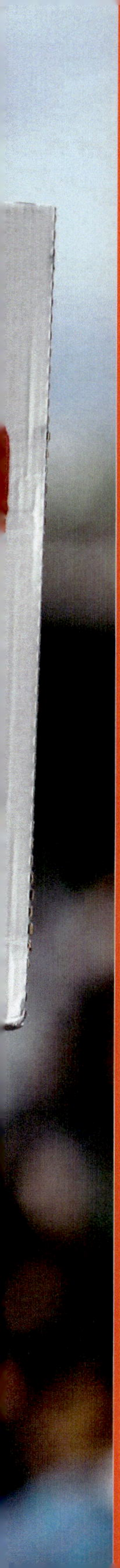



# A Worldwide Effort

Because the problems associated with global warming are far-reaching and international in scope, the issue of how to reduce emissions has become a political one for many governments. Some politicians argue that it would cost too much money to change from fossil fuels to other energy sources. Others want more proof about the detrimental effects of global warming; they say that there is not a **consensus** among scientists. However, in 2007,

**OPPOSITE:** Swedish teenage activist Greta Thunberg inspired millions of people to join a global climate strike in September 2019.

TAKEAWAY

## The human effect on the environment has been recognized as cause for worldwide concern since the 1970s.

the International Panel on Climate Change (IPCC) concluded that the warming that occurred over the previous 50 years could not be explained without taking human activity into account. Most prominent scientists and major scientific organizations have accepted the IPCC's statement as accurate, and more politicians are conceding that climate change is a reality that the world must face.

The human effect on the environment has been recognized as cause for worldwide concern since the 1970s. At the first World Climate Conference in 1979, many scientists warned that human activity could cause climate change, and countries realized then that reducing greenhouse gas emissions would have to be a global effort in order to make a difference. Since 1992, 197 nations have signed the United Nations Framework Convention on Climate Change (UNFCCC), an international treaty that binds nations to work together to combat global warming.

In 1997, delegates from 170 nations met in Kyoto, Japan, to discuss the climate change problem and signed an addition to the UNFCCC called the Kyoto Protocol. According to the Kyoto Protocol, by the year 2012, industrialized countries would reduce

## Alternate Energy

Biogas is an alternative source of energy that is produced from the decomposition of organic material—matter that has come from a once-living organism—in the absence of oxygen. Rotting plants, garbage dumps, and animal waste all produce methane, which, when mixed with oxygen, is combustible. Gobar gas is a type of biogas made from cow dung. At the Gobar Gas Research Station in Ajitmal, India, researchers have designed many gobar gas plants for use around the country. At the gas plants, cow manure is thrown into airtight, circular pits. The manure produces methane gas, which is then piped out for use in people's homes.

greenhouse gas emissions by an average of 5.2 percent below 1990 levels. Countries still in development, such as China, India, and many African nations, would not have to meet the same targets because doing so could hinder the developing nations' growth. The U.S., the world's top greenhouse gas producer, supported the Kyoto Protocol in 1997 but never **ratified** it, explaining in 2001 that adhering to such tight restrictions would hurt the U.S. economy and that the terms were unfair because developing countries did not have to follow the same rules.

Many nations were discouraged by the U.S.'s rejection of the Kyoto Protocol, but the efforts to stop global warming did not end there. Delegates from more than 190 countries, including the U.S., continue to meet under the UNFCCC to propose international strategies for dealing with greenhouse gas emissions. With the Paris Agreement of 2015, parties to the UNFCCC committed to "strengthen the global response to the threat of climate change by keeping a global temperature rise this century well below 2 degrees Celsius above pre-industrial levels."

Although international agreements represent a significant step toward slowing global warming, scientists say it is not enough. In order to stabilize atmospheric levels of greenhouse gases by the middle of the 21st century, we must reduce emissions to 60 percent below 2005 levels. The greatest challenge in reducing emissions is to cut our dependency on fossil fuels by using clean, renewable energy sources. Wind, solar, and hydro (water) power are sources of clean energy—energy that does not emit greenhouse gases or waste. Many countries such as Germany, Spain, and Denmark are already using wind

TAKEAWAY

When sunlight hits the cell, the silicon absorbs the sun's energy, and electrical charges move between the silicon layers to produce an electric current.

energy as a major source of power; even in the U.S., giant wind turbines have been placed in exceptionally windy areas such as the Great Plains. Wind spins the blades on the turbines, which are connected to generators that create electricity.

Hydropower works in much the same way. Ocean waves and waterfalls hold large amounts of energy. Waterwheels can be placed in areas with water movement. The movement of water spins the wheels and generates electricity. Another powerful source of clean energy is the sun. The sun's energy is captured in solar panels that

# Shrinking Footprints

A good way to help curb global warming is to reduce your "carbon footprint." A carbon footprint is the amount of greenhouse gases produced by burning fossil fuels to support your lifestyle. You can reduce your carbon footprint by making simple changes around your home. Exchanging a standard shower head for a low-flow model will reduce the amount of energy used to heat the water. Replacing all regular light bulbs with compact fluorescents will use about four-fifths less energy and save on electricity costs. Drying clothes on a clothesline is also better for the environment, and it will help reduce your footprint, too.

TAKEAWAY

**To produce energy, atoms of the metal uranium are split in a process called nuclear fission.**

are placed on homes, power plants, and other buildings. Inside the panels are solar cells containing thin layers of **silicon**. When sunlight hits the cell, the silicon absorbs the sun's energy, and electrical charges move between the silicon layers to produce an electric current.

Another alternative to fossil fuels is **nuclear** energy, since nuclear power plants do not release greenhouse gases. France receives almost three-fourths of its energy from nuclear power. However, the development of nuclear power plants is a controversial issue for many reasons, not the least of which is that the process is potentially dangerous to humans and the environment. To produce energy, atoms of the metal uranium are split in a process called nuclear fission. When the atom is split, small particles called neutrons are released and bump against other atoms, splitting them and resulting in a chain reaction of nuclear fission. If the chain reaction is not tightly controlled, explosions can occur. Fission releases heat that is used to boil water, which then spins turbines to generate electricity. But **radiation** is also released during fission, and if humans are exposed to high levels of radiation, they can develop cancer and other illnesses.

Moving away from our dependence on fossil fuels is a long process that may take many years. Governments are realizing they need to offer financial **incentives** or penalties in order to speed up the necessary changes. Some have considered issuing a carbon tax, which would charge a fee to companies and factories that use fossil

**OPPOSITE** A variety of organizations have enacted initiatives to reduce carbon emissions.

fuels. Another viable option is carbon trading, in which a cap is placed on emissions, and companies hold a set number of emission allowances, or credits, which cannot exceed the cap. Companies whose emissions go beyond the cap must buy credits from companies with emissions lower than the cap. Therefore, companies with lower emissions can benefit financially.

Transportation uses much of the world's oil supply. As the stock of oil continues to be depleted, and as Earth heats up, automakers are designing more fuel-efficient cars or cars that run on biofuels, such as ethanol. Biofuels are derived from plants and are considered a renewable energy source. Vehicles powered solely by biofuels do not emit carbon dioxide. Another popular alternative is the hybrid car. Hybrids combine a fuel-powered motor with an electric one. The electric motor powers the car

**OPPOSITE** **Improved and lower-cost technologies have led to a resurgence in the use of solar energy.**

at low speeds, and the action of the brakes recharges the motor. While hybrids do use gasoline, they don't need to use nearly as much as traditional cars.

As researchers and governments develop ways to combat global warming, individuals can also play a part in the fight for the earth's life. Turning off lights, appliances, televisions, and computers when they are not in use can help reduce electricity and the greenhouse gases emitted due to its production. Making fewer trips in the car and taking public transportation whenever possible can unclog roadways and decrease air pollution caused by carbon emissions. Once people realize that making even small changes in their lifestyles may help save the entire planet, the threats of global warming will be minimized, and the earth will stand a better chance of remaining a thriving, livable planet fit for many future generations.

# Selected Bibliography


DiMento, Joseph, and Pamela Doughman, eds. *Climate Change: What It Means for Us, Our Children, and Our Grandchildren*. Cambridge, Mass.: Massachusetts Institute of Technology Press, 2007.

Flannery, Tim. *The Weather Makers: How Man Is Changing the Climate and What It Means for Life on Earth*. New York: Atlantic Monthly Press, 2005.

Houghton, John. *Global Warming: The Complete Briefing*. 3rd edition. New York: Cambridge University Press, 2004.

Krupp, Fred, and Miriam Horn. *Earth: The Sequel; the Race to Reinvent Energy and Stop Global Warming*. New York: W. W. Norton, 2008.

Lomborg, Bjorn. *Cool It: The Skeptical Environmentalist's Guide to Global Warming*. New York: Alfred A. Knopf, 2007.

Romm, Joseph J. *Hell and High Water: Global Warming—The Solution and the Politics—and What We Should Do*. New York: HarperCollins, 2007.

U.S. Environmental Protection Agency. "Climate Change." http://www.epa.gov/climatechange/index.html.

Weart, Spencer R. *The Discovery of Global Warming*. Cambridge, Mass.: Harvard University Press, 2003.

# Endnotes

**advocate** a person who publicly supports and recommends an idea or plan

**altitudes** heights of objects in relation to sea level

**atmosphere** the layer of gases that surrounds Earth

**consensus** a general agreement among all involved

**correlation** a mutual relationship or connection between two or more things

**crude oil** oil in its natural state, before it is made into gasoline

**developing countries** the poorest countries of the world, which are generally characterized by a lack of health care, nutrition, education, and industry; most developing countries are in Africa, Asia, and Latin America

**dikes** long, high walls built to hold back water and prevent flooding

**economic** relating to the economy, the social system that involves production and consumption of a country's wealth, goods, and services

**food chain** a system in nature in which living things are dependent on each other for food

**incentives** things, such as money, that motivate people to take specific actions

**Industrial Revolution** a period during the late 18th and early 19th centuries in Europe and the U.S., marked by a shift from economies based on agriculture and handicraft to ones dominated by mechanized production in factories

**irrigation** the distribution of water to land or crops to help plant growth

**latitudes** regions referenced by their temperature and distance north or south from the equator

**malaria** a sometimes fatal disease caused by parasites that is spread by mosquitoes to animals and people

**migrate** to move from one region or habitat to another, especially when the seasons change

**nuclear** having to do with energy created when atoms are split or joined together

**organic** derived from or relating to living matter

**ozone** a toxic gas made up of three oxygen atoms; in the stratosphere, the ozone layer contains a high concentration of ozone, which helps to protect Earth from the sun's harmful rays

**phytoplankton** microscopic plants that drift along near the surface of large bodies of water

**radiation** the emission of radioactive materials, which release energy in the form of heat and light

**ratified** — formally approved a treaty or agreement so that it could take effect

**respiration** — the process by which animals take in oxygen and release carbon dioxide

**silicon** — a chemical element found in sand and rocks, often used in making electronics, glass, and cement

**synthetic** — made artificially through chemical processes, especially to imitate a natural product

# Websites

### JPL: Teach
*https://www.jpl.nasa.gov/edu/teach/tag/search/Climate+Change*

Find activities, resources, and other materials about science topics, including climate change.

### NOAA: Teaching Climate
*https://www.climate.gov/teaching*

From videos to experiments, learn more about climate and energy.

Note: Every effort has been made to ensure that any websites listed above were active at the time of publication. However, because of the nature of the Internet, it is impossible to guarantee that these sites will remain active indefinitely or that their contents will not be altered.

CLIMATE
ACTION
NOW

# Index


Africa 63
animals 9, 15, 17, 18, 40, 48, 55, 56
  coral 40
  golden toads 48
  polar bears 55
  and respiration 17
Arrhenius, Svante 21
atmosphere 9, 15, 17, 18, 20, 21, 24, 25, 26, 32, 35, 36, 39, 52, 64
carbon cycle 17, 18, 21, 24, 32, 33, 34
  carbon sinks 18, 24, 32, 33, 34
Central America 48, 52
China 31, 32, 47, 63
clean energy sources
  *See* renewable energy
climatologists 14, 25, 35, 52
  and ice core studies 25
Engelman, Robert 10
Europe 49, 53, 64, 68
fossil fuels 18, 20, 21, 29, 31, 32, 38, 41, 42, 54, 59, 64, 66, 68, 69, 71, 72
  burning of 20, 21, 29, 32, 66
  and clean coal technology 54
  coal 18, 31
  formation of 18
  incentives against using 69, 71
  natural gas 18, 31
  oil 18, 31, 38, 71
  and transportation 31, 41, 42, 71, 72
    fuel-efficient vehicles 71, 72
Fourier, Jean-Baptiste 18
global temperatures 9, 10, 14, 18, 20, 26, 47, 53, 56, 64
  and desertification 53
  and heat waves 53
global warming causes 10, 11, 14, 19, 21, 24, 29, 34, 35, 45, 59, 60, 61, 64, 72
  deforestation 19, 34
  emissions 21, 59, 61, 64
  human activity 10, 11, 14, 21, 24, 29, 60, 72
global warming effects 10, 26, 45, 46–47, 49, 52, 55, 56, 59
  on human health 56
  melting ice 10, 26, 45, 46–47, 55
    albedo effect 46–47
    Larsen B Ice Shelf 46
  rising and warmer seas 10, 26, 47, 49
  stronger storms 10, 26, 52
  water stress 49
Gore, Al 10, 50
greenhouse gases 9, 18, 20, 21, 22, 24, 25, 26, 29, 31, 32, 33, 34–35, 36, 38, 39, 41, 42, 45, 54, 61, 63, 64, 66, 68, 72
  carbon dioxide 9, 18, 20, 21, 24, 25, 29, 31, 32, 33, 34–35, 42, 54
  greenhouse effect 9, 18, 22, 24, 39
  methane 20, 35, 38
  nitrous oxide 20, 39, 41, 42
  ozone 20, 36
  water vapor 20, 22
habitats 10, 40, 48, 55, 56
ice ages 13–14
India 62, 63
  Gobar Gas Research Station 62
Industrial Revolution 20, 21, 39
International Panel on Climate Change (IPCC) 60
international treaties 61, 63, 64
  Kyoto Protocol 61, 63
  Paris Agreement 64
  United Nations Framework Convention on Climate Change (UNFCCC) 61, 63
Lackner, Claire and Klaus 33
  synthetic trees 33
lifestyle changes 66, 72
  carbon footprints 66
nuclear energy 68–69
oxygen cycle 17
plant life 9, 15, 17, 18, 32, 55, 71
  and photosynthesis 15, 17, 32
renewable energy sources 64–65, 68, 71
  biofuels 71
  solar power 64, 65, 68
  water (hydro) power 64, 65
  wind power 64–65
South America 47
Union of Concerned Scientists (UCS) 19
United States 10, 31, 32, 46, 47, 50, 52, 63, 64–65
World Climate Conference 60
world population 39, 41